CBRN FRAMEWORK AGREEMENT MANUAL

Published by CBRNE Ltd

Title:	CBRN Framework Agreement Manual	
Date:	August 30, 2014	
Author(s):	John Astbury CBE, Dominic Kelly	CBRNE Ltd

This project has received funding from the European Community's Seventh Framework Programme. The views expressed in this document are purely those of the writer and may not in any circumstances be regarded as stating an official position of the European Community.

Front Cover Design by: Carolyn Smith BA (Hons) Ind Des MFA - CBRNE Ltd Design Director

Contents

Glossary of Terms

Term	Definition
CBRN TERRORISM	The deliberate release of CBRN materials (or where a cause remains undetermined).
TOXIC INDUSTRIAL CHEMICALS	Potentially hazardous chemicals used in industrial processes
COLD ZONE	The area furthest away from the source of contamination
CONTAMINATED SITE	Any building, infrastructure (including transport) and open environment
DECONTAMINTION PROCESS	The scientific or technical element of the service i.e. the actual technology or application. The process (such as fumigation, chemical reaction or surface removal) by which contaminants are removed or destroyed
DECONTAMINATION SERVICE	The provision, deployment and management of environmental sampling, decontamination or contaminated-waste management processes from inception to completion
EMERGENCY RESPONSE & RECOVERY	Post Incident and recovery (remediation phase)
FRAMEWORK AGREEMENT	General term for agreements with suppliers which set out terms and conditions under which specific purchases (call-offs) can be made throughout the term of the agreement
AUTHORITY	The government service with responsibility for decontamination services
HAZMAT	Accidental releases of hazardous, industrial, chemical, biological or radiological materials
HOT ZONE	The area closest to the source of contamination
OJEU	Official Journal of the European Union
KPI	Key Performance Indicators
METABOLITE	A chemical compound produced when a body or other living organism breaks down (metabolises) another chemical or substance
OPERATIVE	Employee assigned by the specialist supplier(s)/specialist supplier(s) to carry out specific task(s) or a person(s) who operates under the authority of the AUTHORITY
PPE	Personal Protective Equipment

POTENTIAL SPECIALIST SUPPLIER	A Specialist Supplier is an organisation which is bidding for inclusion on the framework
RESPONSIBLE AUTHORITY (RA)	The responsible authority may be specified by statute or, in the case of a private body or company, may be the owner of the building, location or assets affected by a CBRN incident
RPE	Respiratory Protection Equipment
SPECIALIST SUPPLIER	A supplier of decontamination service which has been admitted to the AUTHORITY Framework Agreement
SPORE STRIP	A strip of material containing a known quantity (within determined limits of accuracy) of active spores of a bacterium, which can be placed within an area to be decontaminated and can be analysed after the event to assess the effectiveness of the distribution of fumigant
STAKEHOLDER	A person who has an interest in an outcome or activity because of an effect it may have on them or their organisation
WARM ZONE	The area between the HOT and COLD Zones

1. Executive Summary

Following the discovery and notification of a potential Chemical, Biological and Radiological (CBR) incident and subsequent site assessment and initial sampling by a member state's emergency service, the nature and extent of contamination needs to be assessed. At this assessment stage it may be necessary to call upon the services of specialist suppliers. Specialist suppliers may also be needed at the decontamination verification and waste management (see Work Package 5 Deliverable 5.12: "Remediation Plans and Templates" and D5.11 "Temporary Waste Storage Facilities" for related guidance) stages of an incident. Currently, in many member states, arranging the contract between specialist suppliers and the Responsible Authority (RA) is carried out at the time services are required. This process can lead to delays in the timely provision of services at the contaminated site. In addition prior knowledge of possible requirements and characteristics of a particular event could speed the decision making process and identify the necessary suppliers within a Framework Agreement..

To reduce time spent on contract procurement and negotiation at the outset of an incident member states may wish to introduce pre-arranged framework agreements procured under EU rules as the basis for contracts with RAs. Whilst the existence of framework agreements does not preclude the RA from contracting other specialist suppliers outside of the framework agreement, the agreements nevertheless provide a sound basis for contract negotiation.

Deliverable 7.3 under Work Package 7 (D7.3/WP7) provides a basis for the procurement of framework agreements in the form of a FRAMEWORK Agreement Manual.

2. Introduction

This Manual sets out the service and technical requirements for the procurement of specialist CBRN Services which can be used by the government body responsible for CBRN incidents within EU Member States, referred to hereinafter in this Manual as the "AUTHORITY". The Manual should be considered as a *Framework Agreement Manual* (hereinafter referred to as the "Manual") and is based upon tried and tested processes, protocols and best practice employed by the UK Government Decontamination Service (GDS)[1]. *Information and Guidance toward the completion of a framework agreement is highlighted as is this text. Text that is not highlighted can be used as stated within a framework agreement Statement of Requirements or the Framework Agreement itself* or used to populate a contract between the specialist suppliers and the RA.

---

[1] Project PRACTICE acknowledges the contribution made to this Manual by the UK Decontamination Service (GDS).

3. Statement of Requirements

The European Union (EU) Public Sector Procurement Directive 2004/18/EC[2] include a provision (Article 32) on framework agreements. The Utilities Procurement Directive[3] and Regulations[4] include this provision at Article 14 and Regulation 19 respectively. This Manual offers guidance to practitioners on the implications of the provision on framework agreements. The Regulations define a framework agreement as:

"an agreement or other arrangement between one or more contracting authorities and one or more economic operators which establishes the terms (in particular the terms as to price and, where appropriate, quantity) under which the economic operator will enter into one or more contracts with a contracting authority in the period during which the framework agreement applies".

In other words, a framework agreement is a general term for agreements with suppliers that set out terms and conditions under which specific purchases (call-offs) can be made throughout the term of the agreement. For CBRN incidents this includes equipment, supporting resources, estimated costs, training, and any special requirements.

In most cases a framework agreement itself is not a contract, but the procurement to establish a framework agreement is subject to the EU procurement rules (see Section 11). In a few circumstances it may be the case that the framework agreement itself is a contract in its own right to which the EU procurement rules apply. This would be the case where the agreement places an obligation, in writing, to purchase goods, works or services for pecuniary interest. For this type of agreement, there is no particular problem under the EU rules, as it can be treated in the same way as any other contract. However, the term 'framework agreement' is normally used to cover agreements which are not, themselves, covered by the definition of a contract to which the EU rules apply (though they may create certain contractually binding obligations).

Such agreements set out the terms and conditions for subsequent call-offs but place no obligations, in themselves, on the procurers to buy anything. With this approach, contracts are formed under the Regulations only when goods, works and services are called off under the agreement. The benefit of this is that, because authorities are not tied to the agreements, they are free to use the frameworks when they provide value for money, but to go elsewhere if they do not.

It is for AUTHORITYS in member states to determine levels of security clearance for specialist suppliers, aspects of business continuity (supply chain resilience) and insurance arrangements within a framework agreement.

In the event of a CBRN incident the responsibility for its resolution falls to the local Responsible Authority (RA). The RA can be a local government, private commercial body or any entity

---

[2] Directive 2004/18/EC of the European Parliament and of the Council of 31st March 2004 on the coordination of procedures for the award of public works contracts, public supply contracts and public service contracts. Please see: http://eur-lex.europa.eu/LexUriServ/LexUriServ.do?uri=OJ:L:2004:134:0114:0240:EN:PDF

[3] Directive 2004/17/EC of the European Parliament and of the Council of 31st March 2004 coordinating the procurement procedures of entities operating in the water, energy, transport and postal services sectors. Please see: http://eur-lex.europa.eu/LexUriServ/LexUriServ.do?uri=OJ:L:2004:134:0001:0113:EN:PDF

[4] Directive 2004/17/EC of the European Parliament and of the Council of 31st March 2004 coordinating the procurement procedures of entities operating in the water, energy, transport and postal services sectors. Please see: http://eur-lex.europa.eu/LexUriServ/LexUriServ.do?uri=OJ:L:2004:134:0001:0113:EN:PDF

affected by a CBRN incident. The RA invokes the agreement to provide the services it requires. The relationship between the AUTHORITY and the RA is set out below (Contractual Arrangements).

The headings and content of this Manual are set out to facilitate, and can be used to populate, a specification of specialist services between the AUTHORITY and suppliers of services in whole or part and the Framework Agreement. It can also be used to populate a contract between RAs and specialist suppliers under the framework agreement. A specimen contract is not included within this Manual as the legal processes and applications may differ between member states. The guidance in this Manual is not intended as a substitute for project specific legal advice, which should always be sought by a contracting authority where required.

A diagram to decide whether or not your procurement is a framework agreement is shown at Figure1.

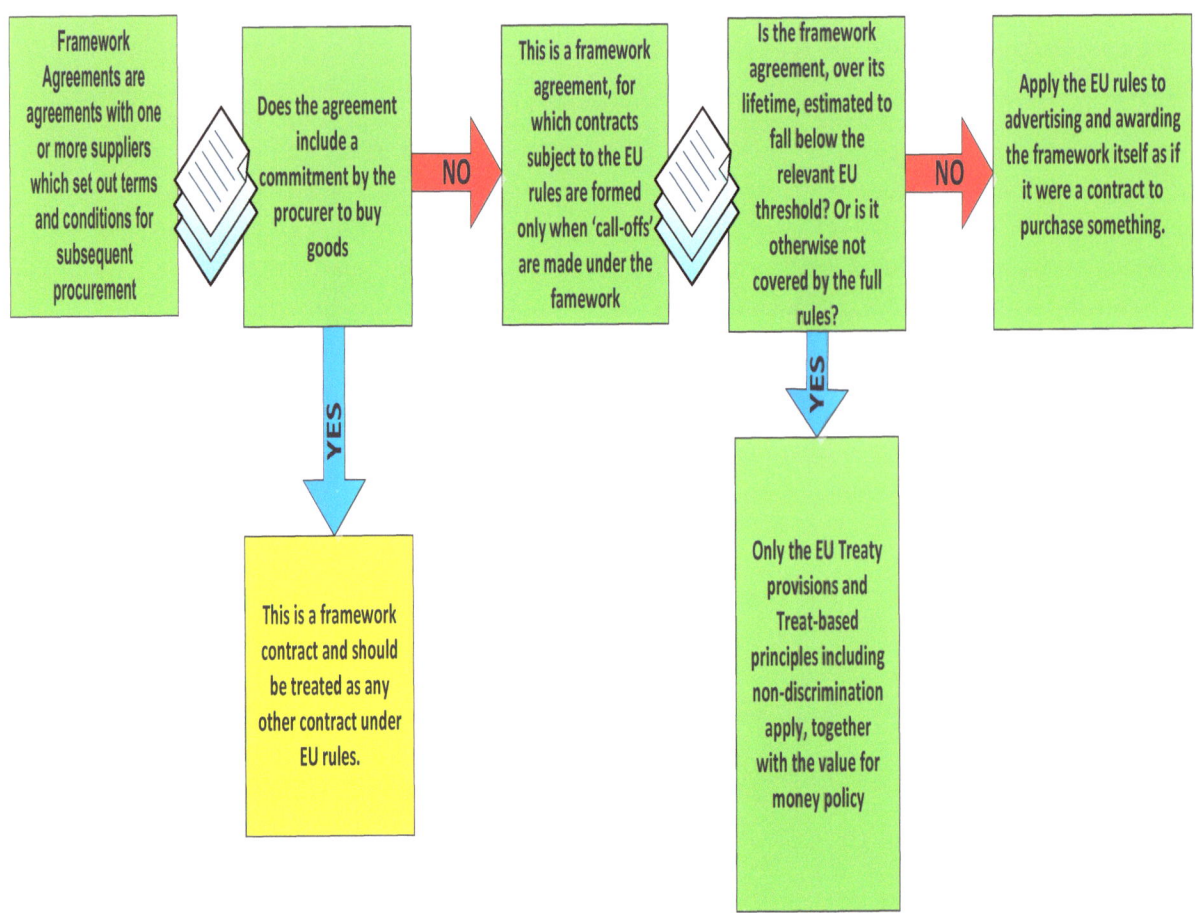

Figure 1: Framework Agreement Decision Diagram

4.   Obligations of Specialist Suppliers

The AUTHORITY requirement is for the specialist suppliers to:

- Supply, deploy and implement a successful on-site sampling, decontamination and waste management service under a contract agreed with an RA to pre-agreed deployment timescales, objectives and standards. These will be set out in the contract for deployment.

- Work closely with the RA at all stages of decontamination, including sampling, planning and carrying out of the decontamination works and the management and transport of wastes.

- Keep costs at a reasonable level by operating a real time cost and scope management process at any incident

- Provide advice on their decontamination service to the RA (and AUTHORITY if requested), as and when requested, working co-operatively in a multidisciplinary team[5] lead by the RA

- Participate in training events and exercises, planning for incidents and other meetings

- Ensure Health & Safety legislation compliance and protection of the environment under any other prevailing legislation as appropriate

---

[5] Team could include government health protection and, environmental bodies, and other government departments or private bodies

5.   Contractual Arrangements

The contract(s) for the provision of the service in any given situation will be between the specialist supplier and the Responsible Authority (RA). The RA, for example, may be a private commercial body or a local government authority. However, the pool of suppliers from which the RA may choose will normally be determined by means of an over-arching framework agreement, or a series of agreements, established by the AUTHORITY following competitive tendering[6]. Additional suppliers may subsequently be added to the framework by further iterations of the procurement exercise.

The framework agreements will be called off by the RA in the event of an incident. The agreement will include details such as terms and conditions, and a schedule of rates, agreed in advance of an incident.

The AUTHORITY should encourage potential specialist suppliers to submit innovative ideas and suggestions relating to the structure of agreements, or the provision of services, that will reduce costs and/or improve the level of service.

The AUTHORITY should develop generic Terms and Conditions (T&Cs) for the guidance of RAs and specialist suppliers. These will be circulated with Invitation to Tender (ITT) documentation. The generic T&Cs should be regarded as specimen T&Cs to facilitate the rapid placement of contracts for decontamination services by RAs. Specific T&Cs for the incident in hand will need to be negotiated with the RA when agreeing the contract for deployment. If the RA cannot reach agreement with the specialist service supplier under the framework agreement, the RA is free to explore and contract other suppliers not covered by the framework agreement (See Figure 2 and Glossary of terms for definitions).

---

[6] A RA is not bound by the framework agreement and is free to choose another specialist supplier as circumstances dictate.

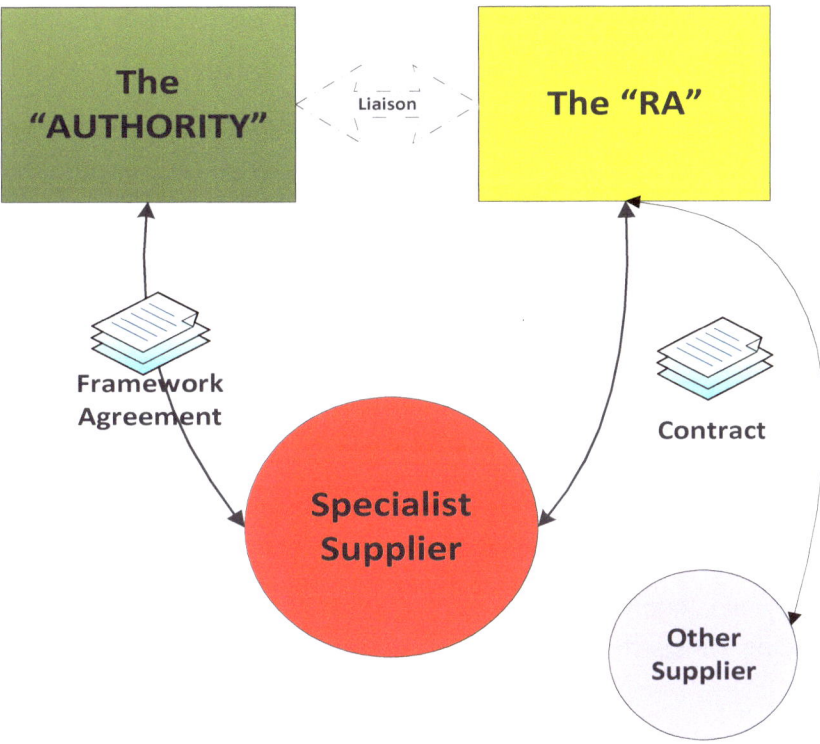

Figure 2: Framework Agreement and Contract Relationship between AUTHORITY, RA, Specialist and Other Suppliers

6. Attendance at CBRN Incidents

The specialist suppliers should:

- Provide access to a competent senior manager appropriate to the incident at an incident or command location as and when specified by the RA

- Provide competent advice on their techniques, systems (and equipment), procedures and processes for the assessment, containment and treatment of the containment

- Work with the RA to remedy the situation until such time as the decontamination objectives have been met and RA decides that the work should cease

The AUTHORITY requirement is for specialist suppliers when requested to deploy to:

- Supply a proposal and agree a contract with a RA in accordance with the terms of the framework agreement

- Provide a detailed plan of the decontamination process, from call out in response to an incident, to the deployment of the service on site, to the successful completion of the decontamination activity. The specialist supplier must ensure that the response times in the plan are kept to a minimum

- Provide and sustain sufficient resources for a successful decontamination service

- Assess and manage all identifiable risks involved in planning for and delivering the service

- Deliver the contracted services at the agreed cost and advise and agree with the RA immediately, any potential changes to scope and cost of the service

6.1   Developing and Supporting Better Preparedness

The AUTHORITY requirement is for specialist suppliers to:

- Develop contingency plans for the decontamination of important buildings, critical national infrastructure (including transport) and areas of open environment, at the request of RA

- Attend meetings and make presentations on the decontamination of CBRN incidents, with a view to enhancing resilience and preparedness

- Participate in training events, and local and national exercises.

6.2   Criteria for the Provision of Specialist CBRN Services

The service supplied shall be:

- Timely: Objectives will be agreed on a case-by-case basis and defined in the contract placed with the RA

- Effective: Decontamination of building structures, surfaces, mechanical and electrical systems, contents, and the wider environment will be to standards agreed and defined in the contract

- Compliant: With appropriate environmental and safety legislation. Codes of Practice, relevant guidelines, or with requirements defined in response to an event by an appropriate authority the RA or the Police

- Considerate: Of the need to preserve as many items or areas within the contaminated zone as possible. Unless otherwise agreed, the service should avoid unnecessary damage to building structures, surfaces, mechanical and electrical systems, contents, other assets (including transport) or the wider environment

- Safe: Deployment of the specialist supplier's own or contracted personnel shall be in full compliance with current national and EU health and safety legislation

- Responsible: The specialist supplier shall rectify any damage to property caused by the negligence of its own or contracted personnel when responding to a CBRN incident

- Measureable: The specialist supplier must undertake to minimise incident costs by operating a real time monitoring and reporting system to identify actual and potential changes to scope and cost to the RA

- Available: The specialist supplier must specify how the service will be maintained and how it will be funded, or, if it has alternate use, how will it be released in response to a RA request. Consideration should be given to importation and other restrictions when sourcing equipment from other member and non-member states

- Sensitive: The specialist supplier needs to be aware of any political sensitivity prevailing at the time when carrying out its work and should avoid any media contact

7. Command and Control

The specialist supplier shall be required to:

- Work within, and to the requirements of, the RA's incident command structure and co-ordination arrangements including supporting public health registers when appropriate

- Comply with instructions and guidelines issued by the RA for dealing with press and other media

- Comply with the prioritisation directed by the RA

- Work with other parties throughout the delivery of the decontamination service until the contaminated site is declared safe for re-use by the RA – this includes any other further decontamination required to achieve pre-agreed decontamination objectives

- Provide the necessary supporting infrastructure for the decontamination service – this includes facilities for the welfare of operatives, whether specialist suppliers or not and their decontamination, for the safe storage of decontaminant(s) and equipment, and other facilities for personnel and equipment as required

8. Standards of Service

The Standards of Service and outputs required from the specialist suppliers will be monitored on a regular basis, by the AUTHORITY.

8.1 Response

The first response to all types of incident will be managed by the appropriate emergency services in the member state. During these times the specialist supplier may be contacted for advice on the decontamination services. The specialist supplier must register the 24/7 contact details of specified individuals with the AUTHORITY for this purpose. The AUTHORITY is expected to host a 24/7 contact telephone number for this purpose.

8.2 Contract Negotiation

The RA will call for proposals from specialist suppliers and prepare the necessary contract for delivery of the decontamination service. At the appropriate time, the RA will direct the contracted specialist supplier to provide the service. The notice required for starting work will be specified in the contract.

> The process for agreeing contracts should be tested during training events and exercises by the AUTHORITY and RAs to ensure that they are fit for purpose.

8.3 Specialist Staff

The framework agreement should specify the specialist supplier's key roles and responsibilities that are needed to provide the decontamination service. Any change to staffing, that would impact the specialist supplier's ability to deliver and sustain the service, must be notified immediately in writing to the AUTHORITY. A revised structure of the response team will need to be submitted.

8.4 Key Performance Indicators

The framework agreement should include agreed performance criteria, measured by Key Performance Indicators (KPIs), which will be used to monitor the performance of the specialist suppliers. Use of these KPIs will be confirmed on deployment by RAs when placing contracts, or with the AUTHORITY when undertaking training events, exercises or reviews of the framework.

> It may be helpful to give an example of a KPI as specialist suppliers may be unfamiliar with the principle. For example, a KPI might set out the response time for a specialist supplier to deploy its resources, equipment or services.

8.5 Technical Resources and Equipment

The framework agreement should detail the technical resources and equipment each specialist supplier should be able to access.

> Evidence of a robust and sustainable supply chain will need to be demonstrated by specialist suppliers, particularly if these resources are to come from outside the AUTHORITY'S jurisdiction, i.e, another member state or non-member state.

8.6 Logistics

Potential specialist suppliers must describe their proposals for storage of equipment when not in use; bulk storage locations; mobilisation procedures; logistical arrangements for shipping equipment to site; detailed arrangements for self-sufficiency on site (catering etc.); and arrangements for removing and transporting all equipment and waste material, once decontamination has been completed successfully. Specialist suppliers will be expected to demonstrate that they can provide a full and appropriate response on initial deployment.

8.7   Notification of Changes

Any material changes to the specialist supplier's ownership, financial position, technology and actual or potential conflicts of interest must be notified to the AUTHORITY immediately, together with notice of any circumstances which will impact upon the framework agreement.

9.   Scientific and Technical Validation

The potential specialist supplier(s) is encouraged to provide innovative specialist services and will be required to:

*   Describe in detail the scientific basis of the tendered decontamination process and cite evidence of independent evaluation and verification of the process wherever possible.

*   Where offered, describe in detail the scientific basis of the tendered sampling/monitoring capability and waste disposal capability, and cite evidence of independent evaluation and verification of such processes.

*   Enable a rigorous technical feasibility study of the tendered processes, by the AUTHORITY and/or RA, or undertaken on their behalf by independent experts and through peer review.

10. Technical Specification of Requirement – Understanding the Role

This section describes the role to be played by specialist suppliers in the response to an incident and the main capabilities desired from them during the recovery phase of the CBRN incident. It is intended to provide an indicative, but not exhaustive, specification of the requirements for delivery of the decontamination service and the context in which that delivery will likely occur (see Table 1 below for the principal roles of specialist suppliers and other key CBRN stakeholders).   Potential specialist suppliers should be invited to offer any further capabilities that they consider may be of interest to the AUTHORITY in this context.

The AUTHORITY	<ul><li>draw up the specification, OJEU notice, for the framework agreement; determine a list of specialist suppliers under the framework agreement for use by RAs</li></ul>
Responsible Authority (RA)	<ul><li>entering into a contract with specialist suppliers under the framework agreement;</li><li>deciding whether to undertake decontamination and, if so;</li><li>how to undertake the decontamination;</li><li>determine clear objectives for the decontamination process.</li></ul>
Emergency Services	All Services: <ul><li>the saving of lives.</li></ul> Police: <ul><li>working closely with other emergency services local authorities and other organisations;</li><li>to secure, and preserve the scene and to control sightseers:</li><li>the investigation of the incident and obtaining and securing of evidence;</li></ul>

	• the collection and distribution of casualty information; • the identification of the deceased on behalf of coroners; • family liaison. Fire: • detection, identification, monitoring and management of hazardous materials; • protecting the environment; • provision of qualified scientific advice in relation to HAZMAT incidents via their scientific advisers • salvage and damage control; and • safety management within the inner cordon. Ambulance/Medical: • to provide treatment, stabilisation and care of those injured at the scene; • to provide appropriate transport, medical staff, equipment and resources; • to establish an effective triage sieve and triage sort system to determine the priority • evacuation needs of those injured and to establish a safe location for casualty clearing, i.e. triage sort area; • to provide a focal point at the incident for Health Services and other medical services; • to provide communication facilities for Health resources at the scene, with direct radio links to hospitals, control facilities and any other agency as required. • to nominate and alert the receiving hospitals from the official list of hospitals to receive those injured and inform other agencies; • to arrange the most appropriate means of transporting those injured to the receiving and specialist hospitals; • to provide decontamination provision for CBRN injured casualties at scene in conjunction with Fire Services
Specialist Suppliers	• entering into a framework agreement with the AUTHORITY; • entering into a contract with the RA for the provision of specialist services for a particular incident; • determining the specific nature of the contaminant(s) and any degradation products • determining the level and extent of contamination • identifying any contamination "hotspots" and precautionary zoning arrangements • detailing the nature of decontaminated materials and surfaces; • undertaking or arranging a detailed post-decontamination sampling and analysis programme; • supporting the emergency services if appropriate specialist services have been defined under the framework agreement

Table 1

Role of Key Stakeholders in CBRN Incidents

It is important that specialist suppliers understand the sequence of events within a CBRN incident and when their services may be required. A suspected CBRN incident would trigger an emergency response, followed by a recovery phase during which the recovery strategy would be developed. The principal phases are set out below and shown in the diagram at Figure 3:

- Discovery and Notification. CBRN contamination is suspected and appropriate authorities are notified. If the evidence is strong, the area will be evacuated and secured. The emergency services will provide an initial response to deal with the crisis (initial rather than crisis) phase of the incident and stabilize the situation. The initial emergency response will normally be led by the member state's Police Service; however this may differ in some member states.

- Site Assessment and Initial Sampling. The nature and extent of contamination on the incident site will be assessed by the emergency services, which will call in appropriate assistance as required. It is not envisaged that specialist suppliers will be required to assist with this initial assessment.

- Defining the Nature and Extent of Contamination. At the start of the recovery phase, command passes to the RA. A more comprehensive monitoring and sampling survey to determine the exact nature and extent of the contamination is likely to be required, and this may be requested from a specialist supplier. This is likely to require a flexible approach to deal with the requirements of real time dynamic risk assessments and a continually developing situation. Transportation of any samples to analytical facilities remote from the contaminated site must comply with all appropriate national and EU legislation.

- Recovery Strategy. The formulation of a detailed strategy for the site will be led by the RA. One or more technical advisory groups will be established to assist and advise as necessary. Technical options for decontamination will be evaluated, set against cost-benefit analysis and hazard evaluation.

- Decontamination Verification. A comprehensive sampling survey will be required to confirm that the decontamination processes have operated as expected and to evaluate the level of any residual contamination. If decontamination has met the required objectives, the area will be released for restoration, refurbishment and eventual re-use. If decontamination objectives have not been met, further decontamination may be required or affected items made safe for disposal as waste.

- Waste Management. Specialist suppliers will be required to manage contaminated wastes by the collection, treatment and transportation to final disposal in accordance with regulations or guidance from RAs. Specialist suppliers will NOT be the owner or consignor of such wastes and will not be responsible for the final disposal of wastes.

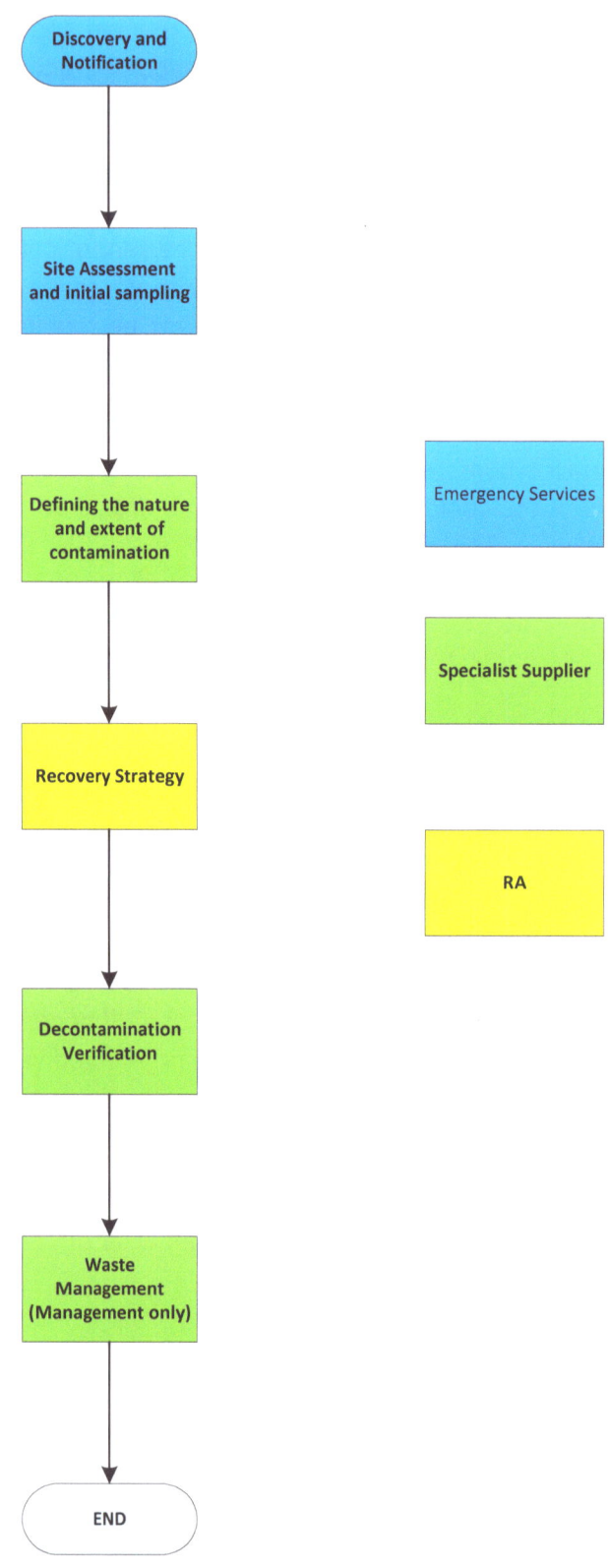

Figure 3: Principal Phases of a CBRN Incident

10.1 Contamination Characterisation and Sampling

Contamination and clearance characterisation is considered across all phases of the CBRN incident as described below.

10.1.1 Emergency Response Phase

Although the AUTHORITY is not a response organisation, where specialist suppliers have capabilities in these areas they may wish to detail them within their response phase, thus allowing the RA the option to utilise these capabilities at the recovery stage and prior to the commencement of decontamination.

10.1.2 Recovery Phase – Initial Contamination Survey

Specialist suppliers may be required to undertake or arrange a detailed sampling and analysis programme to further characterise the contamination. In these circumstances it would be necessary for the RA to ensure that the contaminant was being contained, and to inform the subsequent construction of the recovery strategy.

10.1.3 Recovery Phase – Decontamination Verification

Specialist suppliers may be required to undertake, or arrange, a detailed post-decontamination sampling and analysis programme to demonstrate that the contaminant(s) (including any potentially harmful degradation products), and any decontaminants used in the process, have been removed or destroyed to the pre-agreed toxicity levels appropriate for the intended future use of the site. The sampling procedure will be required to satisfy relevant regulatory criteria and be robust enough to ensure public confidence if the site is to be returned to general use. Verification may be sought by a specialist supplier of sampling services independent from the decontamination specialist supplier.

10.1.4 Method of Sampling and Analysis

Selection of the most suitable sampling and analysis methods will be critical to the success of both pre- and post-decontamination site characterisation. Specialist suppliers will be required to consider the following:

- The type of contaminant (whether chemical, biological or radiological, or a combination thereof)

- The state and stability of the contaminant (whether solid, liquid, gas or vapour particulate or aerosol)

- Any potential breakdown products or metabolites

- The applicability and statistical validity of the proposed sampling methods for the contaminated environment and the different locations and types of materials and surfaces therein

- The impact of varying ambient or climatic conditions

- The number and location of sampling points to ensure the statistical validity of the results

- The efficiency of the sampling method(s) in disclosing entrapped or absorbed contamination

- Limits of detection and quantification

- "Positive challenge" sampling methods (e.g. the placement of spore strips to assess the efficacy of a biological decontamination process)

Specialist suppliers providing this capability will be required to analyse, or arrange the analysis of samples in accordance with good laboratory practice; ensuring auditable sampling procedures are followed. Accredited laboratories only should be used.

10.1.5 How clean is Clean Enough?

Before the contaminated site can be returned to use, the specialist supplier must demonstrate that it has been decontaminated to a level at which it is safe for its intended use. Tolerable levels will be determined on a case by case basis. Where previous attempts have been made to estimate tolerable levels for some highly toxic materials, the levels given are so low as to be extremely challenging to detect. Verification that a decontamination process has achieved tolerable levels of cleanliness is therefore potentially problematic. Specialist suppliers may be required to advise the RAs on achievable levels of detection and to demonstrate they understand what is involved in making such decisions. Normal background levels of contamination may provide a benchmark for establishing fitness for future use.

10.2 Recovery Strategy

10.2.1 Considerations for the Formation of a Recovery Strategy

> The RA will lead the construction of a recovery strategy in consultation with other agencies and key stakeholders. The strategy will include the decontamination process, and related activities such as sampling, analysis and waste management. It is important to note that whilst an individual approach will be required for each situation, generic pre-event recovery strategies may allow some basis for contingency planning. Ultimately, however, each situation will have to be assessed individually.

Specialist suppliers will be a required to advise RAs during the formation of the strategy. Some essential factors that will be considered during this planning process are listed below.

10.2.2 Decontamination Objectives

The RA supported by advice from other agencies and specialist suppliers as appropriate will define clear objectives for the decontamination process. These objectives will be guided by:

- The toxicity and persistence of the contaminant(s) and safe exposure levels where available

- The options for future use of the contaminated site

- The cost effectiveness of achieving particular objectives and the associated risks

10.2.3 Hazard Evaluation

Hazard evaluation is a decision-making process in which the merits of not treating the contaminated site (the "do-nothing" approach) are weighed against the potential benefits and risks from decontamination. However, for a CBRN incident the "do-nothing" approach is unlikely to be appropriate given the complexities surrounding the issue, including public perception and the need to restore confidence. Tolerable exposure levels, to the extent that they are commonly understood, will be the responsibility of the specialist supplier during the hazard evaluation.

10.2.4 Decontamination Cost-Benefit Analysis

The anticipated future use of a contaminated site will influence the type of decontamination methods that will be suitable. If re-use of the site or asset is required, then methods involving demolition or extensive destruction or removal of contaminated materials may be in-appropriate. Correspondingly, the options for re-use will in part be governed by the efficacy of available decontamination technologies and their suitability for the site. Special consideration may be required for key (iconic) buildings or areas. The decisions relating to whether to undertake decontamination and, if so, how to undertake the decontamination will be taken by the RA, which will also lead on the establishment of priorities and determine the order in which areas within the site are to be decontaminated.

10.2.5 Evaluation of Potential Decontamination Processes

Determining the most appropriate decontamination process or combination of processes is central to the development of a successful decontamination strategy. Processes should be appropriate to the particular site and the many potential material/contaminant combinations. The RA will lead the assessment of the suitability of each potential decontamination process or combination of processes on the basis of:

- The effectiveness of the decontamination process and the track record of the potential specialist suppliers of the process(es)
- The applicability of the decontamination process(es) for the contaminated site, the range of materials therein and the prevailing ambient or climatic conditions
- The likely duration of the decontamination activity
- A risk assessment
- The likely environmental impact, including the quantities and nature of wastes generated
- The likely extent of structural or material damage
- Cost

10.3 Technical Requirements upon Deployment after a CBRN incident

The Technical requirements upon deployment following a CBRN incident are described below.

10.3.1 Preparation of a site decontamination plan

The specialist suppliers will be responsible for the production of a comprehensive decontamination plan if they are selected during the evaluation phase (See below and D5.12 "Remediation Plans and

Templates"). This should be submitted to the RA for agreement and incorporation into their recovery strategy. The plan should describe all aspects of the application of the decontamination process to be used, including the quality control and site management procedures to be followed, the equipment and support facilities required (i.e. laboratories, personnel decontamination stations), the methods of waste disposal, worker health and safety precautions and a timetable of the whole process leading to the successful completion of the decontamination activity. All necessary preparation of the site required for successful decontamination must be described in the decontamination plan.

10.3.2 Preparation for Decontamination

The RA and such experts as they specify will agree the decontamination strategy and the specific decontamination plan. The ownership of waste generated from the decontamination process is the responsibility of the RA, but the RA may request advice from the specialist supplier. Specialist suppliers are responsible for their own waste.

10.3.3 Installation of Decontamination Equipment

The selected specialist supplier(s) shall be required to provide, install, commission and operate all equipment necessary for the successful completion of the decontamination activity. This equipment includes that necessary to deliver the decontaminant (or process), to control associated environmental variables such as temperature and relative humidity and to monitor the decontamination process. By arrangement, it may be possible to interface with existing building facilities, such as heating, ventilation and air-conditioning systems where available, to assist the decontamination process. Maintenance of the equipment whilst on site and the decontamination and removal of the equipment on completion of the work will be the responsibility of the specialist supplier.

10.3.4 Supply of Utilities

It cannot be assumed that electrical, water or gas utilities will be available at the site and specialist suppliers should be able to make arrangements for independent provision of these utilities as necessary. Where mains utilities are available, the specialist supplier is responsible for providing any equipment necessary for interface to member state systems (e.g. step-down transformers for electrical equipment or adapters for access to member state hydrants)

10.3.5 Conduct of the Decontamination Process

Specialist suppliers will be required to deliver their capability in accordance with the agreed decontamination plan and the contract with the RA. All consumables, personnel and support facilities necessary to sustain and complete the decontamination process within the contracted timetable are to be provided by the specialist supplier.

The effectiveness of the decontamination process should be monitored by the specialist supplier as necessary.

Any residual decontaminant on the site should be safely removed where necessary on completion of the decontamination process.

10.3.6 Decontamination Verification

A thorough programme of sampling and analysis will be required to determine if the decontamination process has met the pre-defined clearance objectives (See above "How clean is enough?"). Where objectives have been achieved there may be a requirement to repeat the decontamination process, or use an alternative process, after modification of the decontamination plan if necessary.

10.4 Waste Management (See also D5.11 "Criteria for the selection of temporary waste storage facilities")

Specialist suppliers will be required to manage contaminated wastes by the collection, treatment and transportation to final disposal in accordance with regulations or guidance from the RAs. Specialist suppliers will NOT be the owner or consignor of such wastes and will not be responsible for the final disposal of wastes.

Types of waste resulting from assessment and decontamination may include:

- Chemical, biological or radiological wastes

- Bulk solid contaminant

- Bulk liquid contaminant, including liquids absorbed into a solid matrix

- Debris and/or soil and organic wastes

- Soft furnishings and equipment removed from buildings

- Other chemical waste

- Decontaminants (for CBR incidents)

- Chemicals associated with the production or removal of decontaminants

- Waste waters

- Personnel decontamination

- Decontaminant rinse water

- Equipment and consumables (such as personal protective equipment (PPE)) remaining on completion of the decontamination process

- Personal effects belonging to members of the public involved in the incident

The specialist supplier will need to make appropriate arrangements for the minimisation, containment, collection, recording, temporary storage (See D5.11 "Criteria for the selection of temporary waste storage facilities"), transport and safe permanent disposal of such wastes. Where the containment is chemical or biological in nature, there may be benefits in treating contaminated waste at the scene or in dedicated facilities. Disposal will be in accordance to national guidelines according to the types of waste.

11. Health, Safety and Environmental Protection

> Planning and Management
>
> All stages of the decontamination service will require careful planning and management to ensure the health and safety of all involved and protection of the environment from further contamination. There may also be a requirement for the services of independent health, safety and environmental experts to assist with the processes of risk assessment as well as planning and delivery of health and safety (H&S) services.
>
> Risk Assessment
>
> Each activity conducted during the service will require thorough, ongoing and often dynamic risk assessment, an important part of which will be an appraisal of the maximum safe exposure levels for the contaminant procedures and personal protective equipment (PPE) to be selected ensuring personnel are not exposed to levels at or exceeding the safe levels determined. It should also be remembered that many decontaminants themselves represent a considerable hazard and their use must be managed accordingly, and in line with appropriate legislation.

11.1 Health and Safety of Employees

Specialist suppliers are expected to fulfil the following health and safety requirements for their employees:

- The ability to manage a safe site
- Appropriate chemical, biological or radiological hazard awareness training
- Appropriate vaccination and/or prophylactics status (for a biological incident)
- Provision of medical surveillance and support
- Health physics monitoring (for a radiological incident)
- Provision of necessary PPE/RPE
- Provision of personnel trained and experienced in the use of PPE/RPE
- Familiarity with the establishment and management of hot, warm and cold zones for contamination control
- Provision of personnel decontamination equipment
- Personnel decontamination and undressing teams to staff the contaminated zone transition points
- The necessary permissions and licenses to carry, handle, store and use dangerous articles and substances and the ability to do so safely and securely
- Self-sufficient catering facilities to ensure the wellbeing of all personnel
- Maintain appropriate employer and third party insurance for the type of work concerned

12. Goods, Services and Prices

12.1 Schedule of Pricing

Specialist suppliers must provide a schedule of pricing for the service within the framework agreement. The pricing schedule set out in the agreement may however be refined between the supplier and the RA for a particular incident.

12.2 Maintaining a Capability

Any costs associated with maintaining a capability must be specified by specialist suppliers in detail within the schedule – note that retainers will not be paid separately or directly

13. Training Events and Exercises

The framework should include provision for the engagement of specialist suppliers at minor training events with RAs and major national exercises as directed by the AUTHORITY during the life of the contract. Such events and exercises will test, among other things, the call out arrangements of specialist suppliers and deployment of equipment (including KPIs). The contractual arrangements for training events and exercises will be as provided under the framework agreement.

It is the responsibility of specialist suppliers to ensure that they retain their competence through their own training events and exercises outside of the framework agreement. The AUTHORITY should be informed immediately if there is any change to the specialist supplier's capability that will affect their ability to supply the service required under the framework.

14. Procurement Process

Selection of Process

Public services can be procured under four tender processes:

- The Open Procedure
- The Restricted Procedure
- The Competitive Dialogue Procedure
- The Negotiated Procedure

The Key features of each procedure are set out below.

The Open Procedure

The open procedure does not include a pre-qualification stage (see Section 11.8) and allows any interested party to submit a tender for the contract.

The Restricted Procedure

The restricted procedure allows any interested party to request to participate in the contract tender but only those invited by the public body following a pre-qualification stage may submit a tender. The restricted procedure is th most common for framework agreements.

The Competitive Dialogue Procedure

The competitive dialogue procedure is for use in complex projects where the public body cannot adequately specify its requirements. The public body enters into a dialogue with bidders about their requirements before issuing a final tender. The competitive dialogue is seldom used for framework agreements.

The Negotiated Procedure

This is a flexible procurement procedure under which a contracting authority consults specialist suppliers or suppliers of its choice and negotiates the terms of the contract with them. Prior to 2006, complex projects were procured under the negotiated procedure as services contracts. It is now usual that they should be procured under the competitive dialogue procedure and that the negotiated procedure should be reserved for only the most complex projects. The negotiated procedure may also be used in certain other circumstances, for example, where a procurement that was run using one of the other three procurement procedures has failed.

Process for CBRN Framework Agreement

It is usual for the CBRN framework agreement to be conducted initially under the competitive dialogue procedure, but once it has run for a full contract period and experience of specialist suppliers has been gained, the re-letting of the framework agreement is more normally conducted under the restricted procedure.

The Tender Document

The contract should be advertised in the Official Journal of the European Union (OJEU) giving brief details of the services required. The advertisement does not contain the full tender specification; this is provided upon request. An example advertisement for a government framework agreement is shown below in Annex I as it could appear on an OJEU web site. The OJEU Contact Notice must:

Make it clear that a framework agreement is being awarded and:

- Include the identities of all the contracting authorities entitled to call-off under the terms of the framework agreement. The authorities can be individually named, or a recognisable class of contracting authority may be used – e.g. Central Government Departments, local authorities or health authorities in a particular region etc. It should be noted that European Commission guidance on frameworks indicates that classes of contracting authority should be defined in a manner that enables "immediate identification of the contracting authorities concerned. It is not considered to be sufficiently precise to refer to all contracting authorities in a particular region".
- State the length of the framework agreement. It will be a maximum of four years "except in exceptional circumstances, in particular, circumstances relating to the subject of the framework agreement". It is understood that a longer duration could be justified in order to ensure effective competition under the framework agreement if four years would not be sufficient to provide a return on investment.

- Include the estimated total value of the goods, works or services for which call-offs are to be placed and, so far as is possible, the value and frequency of the call-offs to be awarded under the agreement. This is necessary in order for providers to be able to gauge the likely values involved and to provide a figure for the framework overall which, as with other contracts, should not normally be exceeded without a new competition taking place.

Once the OJEU Contract Notice has been despatched, the authorities setting up the framework agreements should follow the rules for all phases of the procurement process covered by the Regulations.

Prequalification Questionnaire (PQQ)

A PQQ should be sent to all those who have expressed an interest in supplying the AUTHORITY with services under the framework agreement following an OJEU tender notice. The PQQ is used to select a shortlist of bidders out of those who expressed an interest. Those bidders who are successful at the pre-qualification stage will then be invited to tender (in the restricted procedure) or invited to negotiate (negotiated procedure), or to participate in dialogue (competitive dialogue procedure).

There are strict limits on the questions which can be included at the PQQ stage. Specialist suppliers should only be asked for information which allows the contracting authority to assess:

- Whether the supplier should or may be rejected (reasons include fraud, bribery, and insolvency
- The economic and financial standing of the supplier and
- The technical and professional ability of the supplier

It is not permissible at PQQ stage to ask how the supplier will approach the particular requirement, or any questions about the supplier's pricing. Contracting authorities should only request information which is to be used as part of the selection process.

A PQQ template is shown at Annex II to this Manual. A PQQ may not be applicable in all member states. Note that the PQQ document is only a template and will require tailoring to suit the specific circumstances of the procurement

15. Tender Specification

The tender specification provides the full range of services required by the AUTHORITY and should set out in detail how the services are to be tendered. Note that the services are divided into 'Lots'. Each Lot covers a specific aspect of the CBRN service required. Potential specialist suppliers can bid for any of the Lots, all of them or a mixture of the Lots as they wish. The AUTHORITY will decide on the competence of each specialist supplier to provide the service for which they have bid. Tendering in this way allows for the widest possible response from specialist suppliers to cover all service requirements.

16. Evaluation of PQQs and Tender Submissions

Whilst the AUTHORITY may wish to process PQQs and tender applications manually (i.e. using AUTHORITY staff competent to do so) they may also wish to consider the use of electronic evaluation tools. These tools can quantify the scoring on both PQQs and any other scoring method the AUTHORITY may wish to use for evaluating the tender. There are several applications available on the market.

17. Conclusion

This Manual describes the role of the AUTHORITY, and the service for which the AUTHORITY intends to place framework agreements with specialist suppliers for deployment by RAs.

RAs will be able to call on these agreements by placing contracts with selected suppliers to deal with CBRN incidents. This Manual is not comprehensive in terms of the requirements of the AUTHORITY and RAs in the event of an incident. Each incident will be analysed on a case-by-case basis due to the unique nature of each CBRN incident. Therefore, deployment response will be required to be both flexible and innovative in order to meet service needs.

Figure 3 sets out the procurement process for arriving at a Framework Agreement.

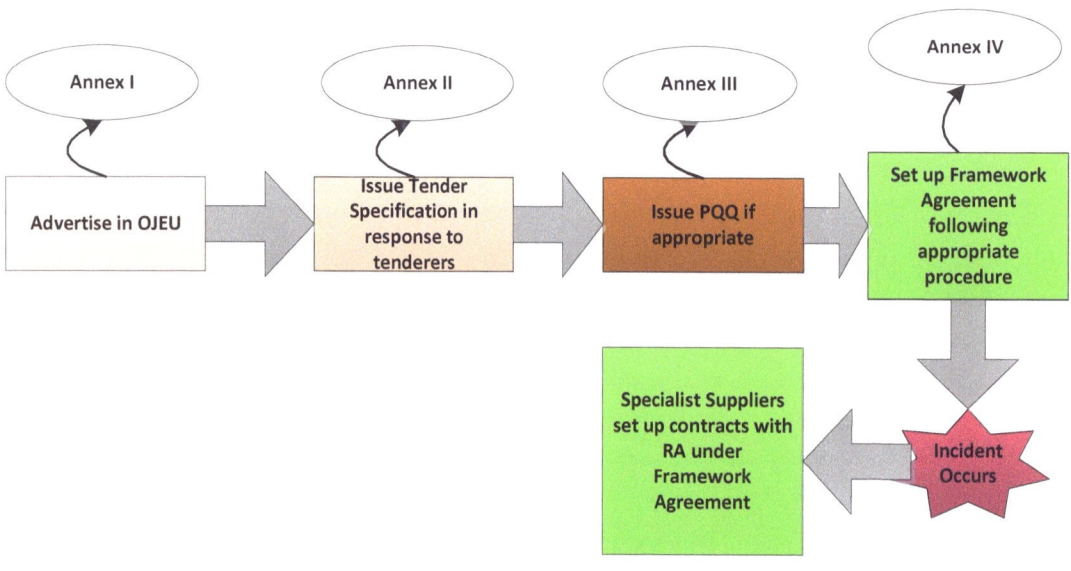

Figure 3

Framework Agreement Procurement Process

Annex I: Example OJEU Advertisement
Annex II: Example Prequalification Questionnaire

• *Annex I Example of an OJEU Advert*

Project Details	
Project Code	e.g. Project_01 (number as appropriate)
Project Title	Government Decontamination Framework Agreement
Opportunity Details	
Project Description	The Authority is seeking to establish a multiple source framework of specialist suppliers (hereafter called the Framework) to provide contingency decontamination and related services in the areas of chemical, biological, radiological and nuclear contamination. The service will include the supply, transport, deployment and implementation of the decontamination and/or related service(s) for use by the Contracting Authority in a CBRN incident. The supplier is expected to provide all necessary equipment required to deliver this service and could include (but is not limited to): raw materials, consumables, protective equipment, detection/monitoring equipment, power generators/fuel, heavy plant, and transportation.
Notes	This is a contingency services framework agreement. Contract spend levels will be determined by occurrences of contamination event. The Authority is unable to guarantee any levels of spend under this framework agreement
Work Category	Services
Procurement Route	EU Restricted
Listing Deadline	14/06/2011 00:00
Contract Start Date	10/02/2012
Contract Duration	4 years
Estimated Value of Contract	Unspecified – see Notes.
Buyer Details	
Organisation	Member state government body
Buyer	Name of government representative (contract contact)
Buyer Email	Email of government representative
Web Link	http://www.government

Annex II Example of a Prequalification Questionnaire

FORM A: Organisation and Contact Details

1.a Enter the full name of the organisation Tendering. If it is a Consortium Tender the Lead Organisation(s) should be entered here. (A Consortium could involve arrangements such as a Lead Organisation with sub-specialist suppliers or a Joint Venture Agreement between 2 (or more) Lead Organisations who have joint and several liability.)	

ORGANISATION DETAILS
(Please provide organisational details for each of the Lead Organisations who will be sharing joint and several liability if a contract is awarded, attaching an Annex where necessary)

1.b Registered office address	2. Company or charity registration number	
	3.. VAT registration number	
	4. Name of immediate parent company	
	5. Name of ultimate parent company	
6. Type of organisation	i) a public limited company. ii) a limited company iii) a limited liability partnership iii) other partnership iv) sole trader v) other (please specify)	

Contact Details	
7. Contact details for enquiries about this PQQ (one contact point only)	
7.a. Name	
7.b. Address	
7.c Post Code	
7.d Country	
7.e Phone	
7.f Mobile	
7.g Email	

8. Type of Arrangement Intended Please indicate which option applies to your intended type of arrangement.	a) Your organisation is bidding to provide the services required itself	
	b) Your organisation is bidding in the role of Lead Specialist supplier and intends to use third parties such as sub-specialist suppliers.	
	c) A consortium is being proposed in the form of a Joint Venture Agreement.	
	d) Other	
Please expand on the type of arrangement proposed in a separate Annex attachment: • Name of Lead organisation(s). Where a Joint Venture is intended, provide details relating to each of the parties who will share joint and several liability. Explain any other type of arrangement being proposed. • Any Parent Company Guarantees that are already in place (or will be available). • Any sub-specialist suppliers proposed at this stage. • A clear chart showing the intended supply chain composition, indicating which member of the supply chain (lead organisation, subspecialist supplier or other) that will be responsible for the different elements of the Procurement.		

QUESTIONS 9.a and 9.b FOR COMPLETION BY NON-UK BUSINESSES ONLY		
9.a Registration with professional body	Is your business registered with the appropriate trade or professional register(s) in the EU member state where it is established (as set out in **Annexes IX A-C of Directive 2004/18/EC**) under the conditions laid down by that member state).	
9.b	Is it a legal requirement in the State /Country where you are established for you to be licensed or a member of a relevant organisation in order to provide the requirement in this procurement? If yes, please provide details of what is required and confirm that you have complied with this.	

FORM B Grounds for mandatory rejection

Important Notice

In some circumstances it is required by law to exclude you from participating further in a procurement process. If you cannot answer 'no' to every question in this section it is very unlikely that your application will be accepted, and you should contact us for advice before completing this form. This will apply in respect of the Lead Organisation(s), Joint Venture Partner(s), sub-specialist suppliers and any other consortium member being proposed.

Please state 'Yes' or 'No' to each question.

Has your organisation or any directors or partner or any other person who has powers of representation, decision or control been convicted of any of the following offences?	
(a) conspiracy within the meaning of section 1 of the Criminal Law Act 1977 where that conspiracy relates to participation in a criminal organisation as defined in Article 2(1) of Council Joint Action 98/733/JHA (as amended);	
(b) corruption within the meaning of section 1 of the Public Bodies Corrupt Practices Act 1889 or section 1 of the Prevention of Corruption Act 1906 (as amended);	
(c) the offence of bribery;	
(d) fraud, where the offence relates to fraud affecting the financial interests of the European Communities as defined by Article 1 of the Convention relating to the protection of the financial interests of the European Union, within the meaning of:	
(i) the offence of cheating the Revenue;	
(ii) the offence of conspiracy to defraud;	
(iii) fraud or theft within the meaning of the Theft Act 1968 and the Theft Act 1978;	
(iv) fraudulent trading within the meaning of section 458 of the Companies Act 1985 or section 993 of the Companies Act 2006;	

(v) defrauding the Customs within the meaning of the Customs and Excise Management Act 1979 and the Value Added Tax Act 1994;	
(vi) an offence in connection with taxation in the European Community within the meaning of section 71 of the Criminal Justice Act 1993; or	
(vii) destroying, defacing or concealing of documents or procuring the extension of a valuable security within the meaning of section 20 of the Theft Act 1968;	
(e) money laundering within the meaning of the Money Laundering Regulations 2003 or Money Laundering Regulations 2007; or	
(f) any other offence within the meaning of Article 45(1) of Directive 2004/18/EC as defined by the national law of any relevant State.	

FORM C Grounds for discretionary rejection

Important Notice

The AUTHORITY is entitled to exclude you from consideration if any of the following apply but may decide to allow you to proceed further. If you cannot answer 'no' to every question it is possible that your application might not be accepted. In the event that any of the following do apply, please set out (in a separate Annex) full details of the relevant incident and any remedial action taken subsequently. The information provided will be taken into account by the Authority in considering whether or not you will be able to proceed any further in respect of this procurement exercise. This will apply in respect of the Lead organisation(s), Joint venture Partner(s), sub-specialist suppliers and any other consortium / supply chain member being proposed.

Please state 'Yes' or 'No' to each question.

Is any of the following true of your organisation?	Answer
(a) being an individual, is bankrupt or has had a receiving order or administration order or bankruptcy restrictions order made against him or has made any composition or arrangement with or for the benefit of his creditors or has not made any conveyance or assignment for the benefit of his creditors or appears unable to pay or to have no reasonable prospect of being able to pay, a debt within the meaning of section 268 of the Insolvency Act 1986, or article 242 of the Insolvency (Northern Ireland) Order 1989, or in Scotland has granted a trust deed for creditors or become otherwise apparently insolvent, or is the subject of a petition presented for sequestration of his estate, or is the subject of any similar procedure under the law of any other state;	
(b) being a partnership constituted under Scots law, has granted a trust deed or become otherwise apparently insolvent, or is the subject of a petition presented for sequestration of its estate; or	
(c) being a company or any other entity within the meaning of section 255 of the Enterprise Act 2002 has passed a resolution or is the subject of an order by the court for the company's winding up otherwise than for the purpose of bona fide reconstruction or amalgamation, or had a receiver, manager or administrator on behalf of a creditor appointed in respect of the company's business or any part thereof or is the subject of similar procedures under the law of any other state?	
Has your organisation	
(a) been convicted of a criminal offence relating to the conduct of your business or profession;	
(b) committed an act of grave misconduct in the course of your business or profession;	
(c) failed to fulfil obligations relating to the payment of social security contributions under the law of any part of the United Kingdom or of the relevant State in which you are established;	
(d) failed to fulfil obligations relating to the payment of taxes under the law of any part of the United Kingdom or of the relevant State in which you are established; or	
e) been guilty of serious misrepresentation in providing any information required of you under Regulation 23 of the Public Contracts Regulations 2006?	

35

	FINANCIAL AND PERFORMANCE MANAGEMENT INFORMATION	
	Please provide one of the following set out below as a separate Annex attachment.	
Indicate which one you have provided typing 'attached' in the relevant box.		
If you are unable to provide option (a), please explain why in your attachment.		
Note: Where a Joint Venture is intended, one of the following options must be provided for each party who will be sharing joint and several liability. Your attachment should be clear on the type of documents being provided for each partner in the Joint Venture Agreement.		
(a) A copy of your audited accounts for the most recent two years		
(b)A statement of your turnover, profit & loss account and cash flow for the most recent year of trading		
(c) A statement of your cash flow forecast for the current year and a bank letter outlining the current cash and credit position		
(d)Alternative means of demonstrating financial status if trading for less than a year		
INSURANCE		
Employer's liability insurance is a legal requirement (except for businesses employing only the owner / close family members) and this should be at least £5 million. Please confirm that you have this in place.		**YES/NO**
PERFORMANCE MANAGEMENT INFORMATION		
(Provide requested information for each Lead Organisation)		
Please provide evidence (as an Annex attachment) that the Lead Organisation(s) have an effective policy and procedure in place for the following areas of Performance Management:		
FORM E Technical and Professional Ability		

FORM E Technical and Professional Ability			
EXPERIENCE AND CONTRACT EXAMPLES			
Please provide details of up to three contracts from either (or both) the public or private sector, that are relevant to the Procurement. Contracts for the supply of goods or services should have been performed during the past three years. Works contracts may be from the past five years. (The customer contact should be prepared to speak to the purchasing organisation to confirm the accuracy of the information provided below if we wish to contact them).			
	Contract 1	Contract 1	Contract 1
Customer Organisation (name):			
Customer contact name, phone number and email			
Contract start date			
Contract completion date			
If you cannot provide at least one example, please briefly explain why (100 words max)			
	Additional Technical Questions **Please provide your response to each of the questions detailed below by attaching as a separate Annex. Your response should be no more than 4 x A4 pages in total.**		
1			
2			
3			
I declare that to the best of my knowledge the answers submitted in this PQQ are correct. I understand that the information will be used in the process to assess my organisation's suitability to be invited to tender for the AUTHORITY'S Procurement and I am signing on behalf of my organisation. I understand that the AUTHORITY may reject this PQQ if there is a failure to answer all relevant questions fully or if I provide false/misleading information.			
FORM COMPLETED BY			
Name:			
Date:			
Signature			